THE NEW SPACE RACE

THE NEW SPACE RACE

CARMEN WILDE

CONTENTS

Introduction to the New Space Race

In the original space race, the United States and the Soviet Union vied for dominance in humanity's final frontier—a contest that ultimately saw the U.S. triumph, the USSR falter, and humanity achieve greatness. Today, the space race has evolved into numerous concurrent stages where commercial, governmental, and non-governmental entities vie for their piece of the cosmic pie. Once, it was a governments-only club. Now, the commercial sector and private innovators play pivotal roles, reshaping an arena that was originally framed by treaties, oversight agencies, and laws, which now strive to keep pace with the rapid commercialization of space technologies and data streams.

In 2018, it seemed like Willy Wonka's factory opened its gates to the cosmos, inviting entrepreneurs to snatch their shiny golden ticket—a harbinger of high-stakes ambitions ranging from orbital outposts to asteroid mining. The new medals of this race? First presence in novel orbits, property rights over resource-rich bodies, and dominion over vital satellite constellations—all competing for the title of the first empire spanning the solar system, aiming for resource bonanzas in High Earth Orbit (HEO), the Earth-Moon Lagrange points (L1 and L2), near-Earth objects (NEOs), and far-off celestial bodies.

Frequently, I pose a question to my audience: What first inspired the journey to the moon? Remarkably, the United States committed a sprawling 4-5% of its annual budget over eight years to send a handful of humans to walk otherworldly terrains. Yet, we did not build lunar colonies or directly progress to Mars. Instead, our gaze shifted. Why did we seemingly abandon spacelines? Today, we're gearing up for a renewed odyssey—a long-term visionary project poised to prove that lunar and interplanetary industrialization is not only feasible but profitable. This resurgence is driven by a new breed of competitors, investors, and billionaires thirsting for an off-planet future. It's my privilege to cheer on this great adventure while advocating for profitable and peaceful exploration. As a historian and futurist well-versed in our past endeavors, my optimism for our future remains steadfast. We can do this. We will do this.

And now, let's enrich the historical background you've outlined:

Historical Background of Space Exploration

Space exploration ignited during the 1960s, primarily spearheaded by the Soviet Union and the United States. This era was punctuated by the iconic moment when Neil Armstrong first set foot on the moon in 1969, fulfilling the U.S.'s monumental goal of landing humans on another celestial body. The intense rivalry between the two superpowers manifested as a dual race for superiority in both space capabilities and high-technology applications for military and civilian advancements.

Over the past decade, a surge of fresh resources has rekindled not only the superpowers' but also other countries' interest in space exploration. These renewed efforts promise an extraordinary future with a plethora of opportunities to enhance life on Earth. However, today's approach to space exploration has shifted towards peaceful endeavors aimed at fostering international cooperation to benefit humanity on our home planet.

At its core, space exploration involves investigating outer space using cutting-edge technology and astronomical expertise. This pursuit significantly expands humanity's capacity for understanding the universe, heralding scientific advancements with far-reaching impacts on industry, the economy, and technological progress. The ripple effects of space exploration offer boundless possibilities, catalyzing economic growth and addressing myriad challenges that humanity faces.

Space technology encompasses a vast array of hardware, procedures, and knowledge pivotal for studying and harnessing the expanse beyond our planet. This includes everything from satellite communications to space travel, driving solutions that resonate profoundly with Earth-centric needs.

CHAPTER 1

Key Players in the Modern Space Race

SpaceX, founded by Silicon Valley entrepreneur Elon Musk in 2002, stands at the forefront of the commercial space sector. Renowned as the largest commercial satellite manufacturer and private space transportation company globally, SpaceX has become synonymous with innovation and ambition. The company has secured hefty contracts for rocket development, design, and manufacture with organizations like NASA, the Department of Defense, and a slew of private customers. Musk's visionary goals extend far beyond near-Earth endeavors; he aims to establish a human colony on Mars, ferrying passengers and construction materials to the red planet. Alongside these grand ambitions, SpaceX is also pioneering space tourism, having already sent a number of affluent adventurers to the International Space Station (ISS) for a hefty price tag exceeding $20 million per ticket. As the cost of space travel continues to decline, the prospect of space hotels and other cosmic ventures becomes increasingly tangible.

The early days of space exploration were marked by a limited rivalry primarily between the USSR and the USA, making cooperation the path of least resistance. Today, the landscape is vastly different. With sufficient knowledge, funding, and investment, it is entirely feasible for new players—nations and even private companies—to venture into space. Over 70 countries currently have active space programs, and emerging space powers like Bermuda, Bahrain, and Kazakhstan are beginning to stake their claims in the cosmos. Between 1958 and 1965, only one country managed to send a satellite into space. Now, over 63 countries have joined the "space race," and this number continues to grow. The 1950s and 60s saw the USA and Russia (then the USSR) as the dominant forces; today, commercial corporations share a prominent role in pushing the boundaries of space exploration.

Government Space Agencies

In the realm of space exploration, national space agencies play a critical role in guiding and executing space initiatives. There are dozens of nations with dedicated space agencies, each with varying missions—some focus solely on satellite operations, while others engage in human spaceflight, space science, exploration, and more. One measure of the global commitment to space has been the widespread adoption of the intergovernmental Agreement on the Rescue of Astronauts, the Return of Astronauts, and the Return of Objects Launched into Outer Space, commonly known as the Rescue Agreement, concluded in 1968. This foundational treaty was further bolstered by the 1979 Agreement Governing the Activities of States on the Moon and Other Celestial Bodies, joined by over 60 countries, though notably absent from this latter agreement are the world's two largest spacefaring nations, the United States and Russia.

A few leading spacefaring nations, through their national space agencies, have achieved remarkable milestones and continue to shape the future of space exploration. These include:

- **NASA (National Aeronautics and Space Administration):** Established in 1958, NASA serves as the United States' principal space exploration agency. From the moon landings of the Apollo era to the present-day Artemis missions aimed at returning humans to the lunar surface, NASA remains at the forefront of space innovation and exploration.

- **Roscosmos (Russian Federal Space Agency):** Russia's Roscosmos, a successor to the Soviet space program, has a rich legacy, including the launch of the first human, Yuri Gagarin, into space. They continue to spearhead a range of missions from Earth observation satellites to deep space exploration.

- **European Space Agency (ESA):** ESA is an intergovernmental organization of 22 member states dedicated to the exploration of space. Collaborative missions with NASA and independent projects such as the ExoMars mission underscore ESA's significant contributions.

- **China National Space Administration (CNSA):** CNSA has rapidly advanced its space capabilities, launching the Chang'e lunar exploration missions and the Tianwen Mars missions, solidifying China's position as a major spacefaring nation.

- **Indian Space Research Organisation (ISRO):** Known for its cost-effective satellite and space mission programs, ISRO has achieved milestones such as the Mars Orbiter Mission (Mangalyaan), becoming the first Asian nation to reach Martian orbit and the only one to do so on its first attempt.

- **Japan Aerospace Exploration Agency (JAXA):** JAXA is recognized for its expertise in asteroid exploration, with missions like Hayabusa and its successor, Hayabusa2, which have successfully returned samples from distant asteroids.

- **French National Centre for Space Studies (CNES):** CNES has been pivotal in developing Europe's space policy and executing various projects, including launch vehicle technology and Earth observation satellites.

These agencies, while often collaborating with commercial enterprises for services and products, play indispensable roles in advancing technology, space science, and exploratory missions. Their commitment ensures the continued pursuit of knowledge and innovation beyond our planetary confines.

CHAPTER 2

Technological Advancements in Space Exploration

Humankind has always been fascinated by the cosmos, and our journey to understand and explore it has led to remarkable achievements and technological advancements. One of the most iconic landmarks in this quest was the lunar landing mission. The Apollo XI mission symbolized human ingenuity, using the limited resources available at the time to achieve the extraordinary feat of landing on the moon. Rockets were able to carry more substantial payloads into space, but guiding them precisely to the moon required groundbreaking innovation. Since then, advancements in propulsion systems have significantly influenced space exploration, facilitating not just short-term missions but also long-term space habitation.

Over the past decade, an array of technological advancements has revolutionized our approach to space exploration. These advancements encompass the launching of both human-crewed and robotic missions as well as the collection and analysis of space data. Innovations in IT and satellite technology, alongside improved communication systems, now allow astronauts and mission controllers to maintain live, instantaneous communication. Here, we will delve into some of the most profound technological advancements in space over recent years, highlighting how humanity is pushing boundaries further into the cosmos.

Rockets and Propulsion Systems

Rockets and propulsion systems are fundamental to the success of any space mission. At their core, rockets are engineered to deliver payloads—whether satellites, spacecraft, or crew—into specific orbits or on interplanetary trajectories. These systems consist of the payload, propellant, and various on-board devices that control trajectory, attitude, and energy supply.

One key concept in propulsion is **propulsive efficiency**, which measures the ability of a propulsion system to use minimal propellant while accelerating heavy payloads to high velocities. This efficiency is typically represented as a dimensionless parameter, defined by the ratio between the increment in mass of the target body and the increment in mass of the spacecraft. This parameter gives insight into the terminal velocity achievable by a unit of thrust without external force. In high-thrust scenarios, propulsive force must counteract gravitational pull, requiring knowledge of the specific propellant properties and internal software controlling thrust.

Rockets remain the only propelled launch systems capable of reaching velocities surpassing the first cosmic velocity (the minimum orbital speed). Their design and operation depend on multiple velocity components:

1. **Specific (internal) terms**: These include the energy, mass, and composition of the propellant, as well as the analytical expression of the thrust producing exhaust gases.
2. **General mechanics terms**: These account for gravitational attraction, aerodynamic loads, thermal effects, and Earth's ro-

tation, providing the dynamic field through which the rocket and propulsion system navigate.

3. **Universal fundamental physics terms**: These relate to the fundamental properties of space and time, influencing how all bodies in the universe move.

Recent advancements in rocket technology have led to significant improvements in propellant efficiency, allowing missions to achieve greater distances and durations. Innovations such as reusable rocket stages, pioneered by companies like SpaceX with their Falcon 9 and Starship rockets, significantly reduce the cost of accessing space. These reusable systems mark a revolutionary shift, making space more accessible and sustainable.

Development in propulsion systems is not limited to chemical rockets. Electric propulsion, which uses electrical energy to ionize propellants and generate thrust, shows promise for long-duration deep space missions. Ion thrusters, for example, offer high efficiency and low thrust, ideal for maintaining satellite orbits and interplanetary travel.

Additionally, advancements in on-board control systems allow for more precise trajectory adjustments and safer mission operations. Autonomous navigation systems, powered by sophisticated algorithms and real-time data processing, enhance the robustness and reliability of space missions.

CHAPTER 3

Commercial Space Industry

Two of the dynamic newcomers reshaping the space industry are SpaceX and Blue Origin, founded by two of the wealthiest individuals on Earth, Elon Musk (worth $192 billion) and Jeff Bezos (worth $189 billion), respectively. Both companies have made significant strides in their space endeavors. In 2020, SpaceX became the first private company to send astronauts to the International Space Station (ISS) and has since been working on developing a spacecraft capable of transporting humans to Mars. Blue Origin, meanwhile, focuses on making space travel more accessible with reusable rockets and aims to democratize space in the long run.

As the United States pivots towards a future driven by commercial space travel, it's crucial to understand how today's civilian space exploration objectives differ from the geopolitical motivations of the superpower race to the moon more than fifty years ago. Sarah Cruddas, author of "The Space Race: The Journey to the Moon and Beyond," emphasizes that today's priorities are markedly different: "Back then it was about geopolitics—the Cold War; it was a way to be seen as more advanced than the enemy, the Soviet Union.

This time it's aligned with capitalism principles, creating a new space economy for Earth."

Private Space Companies

Blue Origin: Jeff Bezos founded Blue Origin with a profound vision: that space can be made accessible to everyone, and that resources from the solar system can benefit all of humanity. From building a road to space to focusing on sustainable and democratized space travel, Blue Origin's mission centers around transforming human life in space. They have been working on developing rockets capable of both suborbital and orbital missions.

Since December 2006, Blue Origin has heavily invested in self-financed development activities. By October 2012, the company reached significant milestones, gaining the capability to carry astronauts. Their successful launches and firings in recent years have ensured that Blue Origin can ferry rockets and space passengers from NASA facilities. Collaborating with the Defense Advanced Research Projects Agency (DARPA) in 2019 bolstered their technological portfolio. Blue Origin continues to push the envelope by innovating technologies designed for cost-effective and efficient space travel—potentially paving the way for sustainable life beyond Earth.

SpaceX: Elon Musk's SpaceX is driven by the ambition to secure humanity's future as a multi-planetary species. Musk envisions colonizing Mars as crucial for both survival and inspiration. SpaceX employs over 5,000 people, designing and launching advanced rockets and spacecraft from their base in Hawthorne, California.

A key mission for SpaceX has always been to make Mars habitable. Their plan, dubbed the Red Dragon mission, involves extracting carbon dioxide from the Martian ground to produce oxygen and methane, ensuring a sustainable return journey to Earth. Besides their Mars-focused initiatives, SpaceX has pioneered the use of reusable rockets, reducing launch costs and making space more accessible. The company has seen remarkable success with their Falcon 9 and Starship rockets, which serve both commercial and government clients.

Both Blue Origin and SpaceX, while driven by unique visions, share a common goal of expanding humanity's reach into space.

Their efforts symbolize the dawn of a new space age—one characterized by commercial innovation and boundless exploration.

CHAPTER 4

International Collaboration in Space Exploration

Humankind's fascination with the cosmos has always been a source of inspiration and cooperation among nations. Sadly, the early momentum for collaborative ventures in space exploration waned due to growing international tensions, shifting political landscapes, ideological conflicts, and military rivalry. Despite the stormy backdrops of Soviet-American rivalry, exemplified by conflicts like the Korean War and the Cuban Missile Crisis, and the lack of trust in regions like the Arctic or the Pacific, specific programs dedicated to cooperative efforts in space exploration managed to persevere.

The history of space exploration is intertwined with political dynamics and diplomatic efforts. Cooperation in space exploration was born during the Cold War, highlighting space as fertile ground for fostering international dialogue. The enormous scientific potential of space provided both the USSR and the USA ample reason to promote peaceful collaboration beyond Earth's confines. The legacy of early space endeavors served as a crucial relic and foundation for ongoing and future peaceful applications of space technology.

The Role of International Space Stations

Space stations are crucial in pushing the boundaries of human spaceflight and fostering international collaboration. These platforms enable us to conduct experiments and research in the unique microgravity environment of low Earth orbit (LEO). By examining the effects of microgravity on cells, plants, and animals, as well as studying the health and performance of astronauts, the International Space Station (ISS) prepares us for future long-duration missions to destinations like the Moon and Mars.

Pioneering space agencies and private companies are developing new capabilities for cargo and crewed missions into deep space. These include advancements in spacecraft reusability and docking systems, along with innovative technologies such as 3D printing. These efforts are paving the way for the next steps in human space exploration.

By the mid-2020s, the first elements of a lunar gateway—a space station intended to orbit the Moon—are expected to be assembled. This gateway will enable lunar surface missions, with astronauts potentially living and working there by 2027. This collaborative project epitomizes the potential of international cooperation in advancing human space exploration.

CHAPTER 5

Manned vs. Unmanned Missions

Robotic missions are characterized by the use of remote-controlled and autonomous reconnaissance devices such as satellites, rovers, and spacecraft. Many of the pivotal discoveries and advancements we have made regarding other planets in our solar system have come from these purely robotic missions. These missions operate without humans on-site, relying on remote communications and decision-making. The advantages of robotic missions include reduced costs, lower risk, and minimized necessity for recovery. They are less constrained by the physiological limits that bind human explorers.

Despite these benefits, humans excel in activities requiring intrinsic intelligence and versatility. Human missions, however, come with significant challenges. For instance, on Mars, human operational performance during extravehicular activities (EVA) is capped at around 4 continuous hours due to high operational costs and the need for precise life-support systems and infrastructure. While astronauts can survive on the Martian surface for months, their external exploration time is limited.

Historically, most space exploration has been accomplished through robotic missions. To date, only the United States and Russia have successfully sent humans into space. One of the primary arguments against human spaceflight is the cost. It is significantly cheaper to send robots or unmanned spacecraft into space. The future of human space exploration beyond the moon remains uncertain due to these financial constraints. The United States' human spaceflight program combines autonomous robotic planetary missions with human missions to leverage the strengths and mitigate the challenges of both approaches.

Benefits and Challenges of Human Spaceflight

NASA's programs have spurred technological innovations that have found applications in various fields. One example is telemedicine and remote robotic surgery, which leverage digital imaging and telemetry. NASA often prioritizes reliability and portability over cost, which sometimes limits the commercial feasibility of these technologies due to their stringent requirements. These parallels between human spaceflight research and public health remain largely unexplored. Notable authorities, including NASA, the U.S. Department of Defense (DOD), and defense leaders from other countries, emphasize that advancements in biotechnology, genomics, and public health, stemming from joint U.S.-Russian investments, can significantly aid in countering biological threats, including terrorism.

Human spaceflight offers numerous benefits beyond increased scientific knowledge, national prestige, and economic gains. These include inspiration, role models, the potential for future planetary settlements, and the validation of new technologies. However, human spaceflight also faces significant challenges, categorized into technical, budgetary, physical health, psychological health, and public opinion. Most of these benefits and challenges are context-dependent. Understanding and addressing these issues is crucial as humanity embarks on the exploration of the final frontier, reaping the benefits and overcoming the hurdles associated with human spaceflight.

CHAPTER 6

Space Tourism: The Future of Travel

The era of space tourism has arrived, opening up the cosmos to those who can afford this extraordinary adventure. It has ignited the imagination of many and turned dreams into reality for a few fortunate individuals. British billionaire Sir Richard Branson, through his company Virgin Galactic, aims to fly up to 3,000 people to space within the first five years of operation. The inaugural flights took place in 2010, bringing astronauts to low Earth orbit and captivating the world with the possibilities of commercial space travel.

Space tourism heralds a new frontier akin to the wild west, where extreme adventure seekers can fulfill their aspirations of becoming astronauts. Early space tourists have already paid substantial sums for this once-in-a-lifetime experience. Earthbound travel companies are eyeing the next big attraction for wealthy travelers eager to explore beyond our atmosphere.

One of the early milestones in private space travel was SpaceShipOne, a privately-designed and financed space plane. The team behind SpaceShipOne won the Ansari X-prize for successfully building a low-cost manned space plane that completed two flights reaching an altitude of 100 km within two weeks.

Current State of Space Tourism

Numerous companies are now competing for a share of the burgeoning space tourism market, offering various experiences based on the level of luxury their clients can afford. SpaceX started flying crewed missions in 2021 with their Falcon 9 Crew Dragon. Under the mission name 'Inspiration4,' sponsored by Shift4Payments, SpaceX achieved the first commercial rocket to space, reaching a higher orbital altitude than the International Space Station. They created a custom-made pod, offering participants a luxurious space travel experience in their all-civilian mission.

Space tourism, a relatively new industry, has seen significant growth in potential revenue and the number of companies entering the field. The success of flights operated by the Russian Space Agency, ROSCOSMOS, and NASA's missions to the International Space Station over the last two decades, coupled with SpaceX's reusable vertical takeoff and landing rocket system (Starship project), has captured the attention of the ultra-wealthy investing in current and future space tourism ventures. Space tourism participants have paid large sums to fly onboard rockets as mission specialists or spaceflight participants. The primary activities in the space tourism industry also include near-space flights using stratospheric balloons and high-altitude suborbital rockets such as those from High-Tech Rocketry Workshops (HTRW).

CHAPTER 7

The Ethics of Space Exploration

Space exploration, while a monumental achievement of human ingenuity, raises various ethical questions about the ownership, use, and impact of space activities. An abandoned satellite, a spent rocket stage, or a used fuel canister has no terrestrial owner. These objects, once they reach Earth's orbit or surface, are considered "relinquished" under the 1967 Outer Space Treaty. However, space objects that remain operable are generally considered the property of the entity responsible for placing them there. The situation becomes complicated when another entity successfully operates and restores functionality to a previously abandoned space object. Many planned national and commercial space activities hinge on the notion of temporary ownership, but this concept is increasingly viewed as unworkable by regulatory authorities.

One of the most contentious issues is whether national security concerns should override the basic norms of the international space legal regime. As more nations and private entities engage in space activities, these ethical tensions will continue to grow, emphasizing the importance of an internationally agreed-upon framework for the ethical governance of space exploration.

Ownership and Exploitation of Space Resources

Advocates argue that developing property rights in space would incentivize a wide array of ventures in the space industry. Sustaining human presence in outer space involves significant costs, which, according to the "province of all mankind" principle outlined in the nonappropriation doctrine, legally belong to the entire international community. The 1998 Declaration on International Cooperation in the Exploration and Use of Outer Space, particularly concerning developing countries, stressed that the costs should be shared and the benefits equitably distributed among member states.

The primary issue to resolve in future space mining endeavors is the ownership and exploitation of space resources. The United States has indicated its intention to encourage private sector involvement in this area through legislative measures like the Commercial Space Launch Act Amendments of 1984 and the International Space Commercialization Act of 1990. These Acts aim to promote non-terrestrial resource utilization for commercial purposes, responding to proposals from entities like Planetary Resources and Deep Space Industries for establishing property rights over asteroid resources.

The foundation for these discussions lies in the principles of responsibility and liability, as highlighted by various international agreements such as the 1972 Space Liability Convention and the 1973 Space Rescue Agreement. The evolving regulatory and ethical landscape will need to address issues of standing, risk-sharing, and dispute resolution to ensure fair and sustainable space exploration practices.

CHAPTER 8

Space Law and Governance

The burgeoning interest in space resource exploitation highlights the need for a robust international legal framework. This may necessitate modifications to existing treaties or the creation of new regulations. In the long-term, establishing comprehensive international space law and governance, focusing on property rights and allocation mechanisms such as transferable development rights or quotas, could form the basis for private ownership of resources on celestial bodies. For now, commercial operators are expected to adhere to established international norms while navigating the complex landscape of space law.

Space technology has advanced rapidly, often outpacing the international legal framework designed to govern space activities. Various United Nations instruments provide basic resolutions, principles, and treaty norms for outer space governance. The 1967 Outer Space Treaty serves as the foundational document, ensuring that space research and exploration benefit all nations. The UN Outer Space Committee remains the primary forum for developing international space law and fostering cooperation. This committee also plays a crucial role in coordinating responses to potential Earth-impacting asteroids.

The Outer Space Treaty

The Outer Space Treaty of 1967, a non-armament agreement, forms the cornerstone of international space law. The treaty outlines that no nation can claim sovereignty or legal rights over any celestial body. While countries can own and operate equipment on these bodies, they cannot claim territorial ownership. Additionally, the treaty addresses liability for damages caused by space activities, holding the launching country responsible for any harm inflicted on celestial bodies or space stations.

One of the treaty's critical elements is the prevention of weaponizing space. It explicitly prohibits placing weapons of mass destruction in space, on celestial bodies, or stationing them in outer space by any means. This regulation is vital for ensuring the peaceful exploration and utilization of space, as it prevents the militarization of a domain intended for scientific and peaceful purposes. The principles enshrined in the Outer Space Treaty continue to guide international efforts in maintaining space as a non-weaponized, cooperative environment for all humanity.

CHAPTER 9

The Future of Space Exploration

The future of space exploration will necessitate both wisdom and global cooperation. Wisdom will safeguard against the exploitation of space and ensure humanity's activities align with ethical and sustainable principles. Cooperation will diminish the high costs associated with space endeavors. Public education will be crucial in fostering a well-informed and supportive public opinion. Educational programs targeting schools, colleges, and adult groups will build understanding and support for the space program, excluding the development and deployment of weapons.

In the coming decades, interest in space exploration will surge. Investments in the industry will continue to grow, surpassing previous peaks annually. The focus on exploration and space settlement expansion will lead to the construction of industrial and research facilities, enabling groundbreaking scientific discoveries. Private industry and enterprise will play a significant role in this future landscape.

Mars Colonization

Mars, the fourth planet in our solar system, is a cold and dry world with high winds that create massive dust storms and frequent dust devils. The planet's rugged terrain includes ancient mountains and craters, with Olympus Mons, the tallest volcano, towering over three times the height of Earth's tallest mountains. Mars has a surface area almost equal to Earth's landmass, providing ample ground for exploration and potential colonization.

Despite evidence suggesting water may have existed on Mars 1.5 billion years ago, no definitive confirmation has been found. For successful habitation, hyperbolic habitats will be essential. These controlled environments will cater to daily needs and ensure astronauts' safety. Establishing a Mars Pioneer outback will involve testing artificial habitats designed to support the first wave of commercially viable human settlers. By prioritizing non-academic benefits and reducing reliance on religious, space agency, military, and university control, a Mars-to-Earth-back-to-Mars space economy can flourish.

We live in an era characterized by exponential growth in scientific research and technological advancements necessary for space travel. While significant progress has been made in unraveling the mysteries of outer space, many challenges remain. Human curiosity and ambition drive us to explore beyond our current boundaries, just as past generations demanded easier access to different regions of our globe. The remarkable advancements in cars and aircraft in the last century have made this possible. Today's generation yearns for commercial space trips and the establishment of colonies on other planets. The extent of our reach will ultimately depend on the level of effort and investment we are willing to commit.